WHAT IS BIOLOGY?

For Curious Little Minds

There are so many exciting things to explore around me. Animals. Plants. Toys.

Wait! Are all these things living?
No.

Can your toys eat and grow?
H_2O
Chemistry Lab
Physics Lab
No. How can I tell if something is living?

The study of Biology can help us answer this.
Biology Lab
What is Bi.ol.o.gy.?

Biology is the study of living things.

It helps explain what living things are, where they came from, what they are made of, and how they grow and change.

There are many topics you can study in Biology. They are called Branches. Can you guess any?
LAB
Animals? Plants?

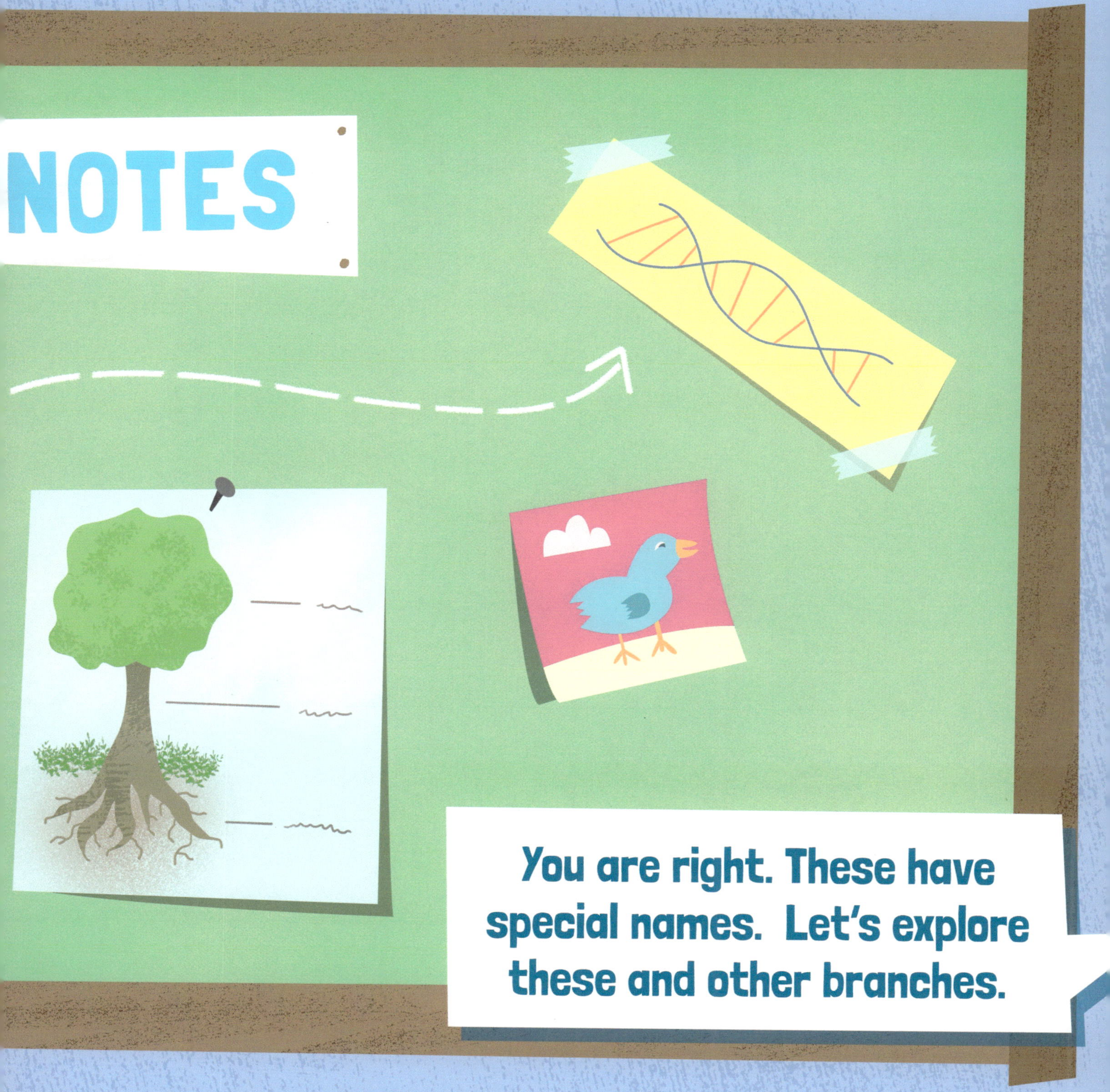
NOTES
You are right. These have special names. Let's explore these and other branches.

Cell Biology

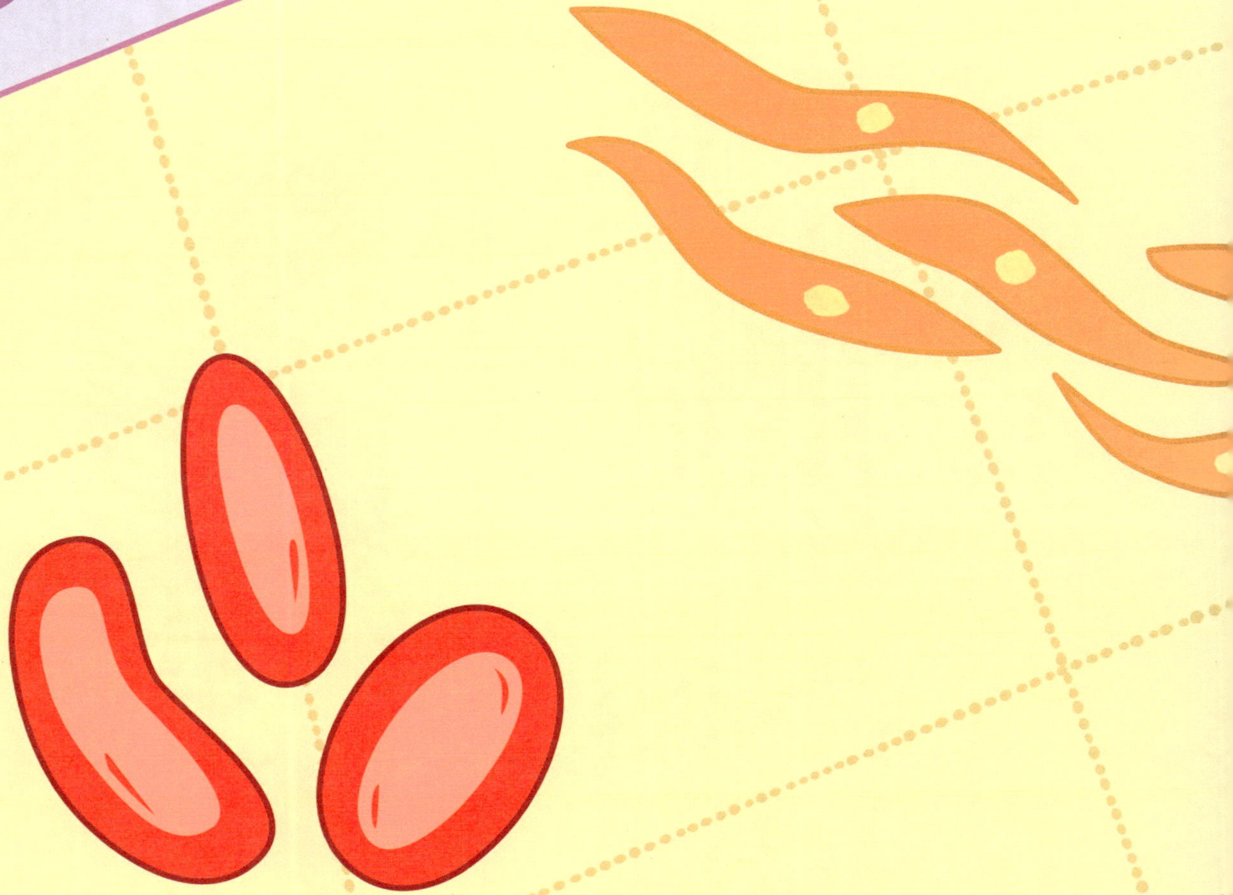

Zoology is the study of animals.

Zoology is the study of animals.

ZOOLOGY

BOTANY

Botany is the study of plants.

Evolution is the study of life and how it changes over long periods of time.
A long time ago...

Evolution
...today!

Genetics is the study of traits and how they are passed on from parents to their children.

GENETICS

Anatomy
Heart
Bones

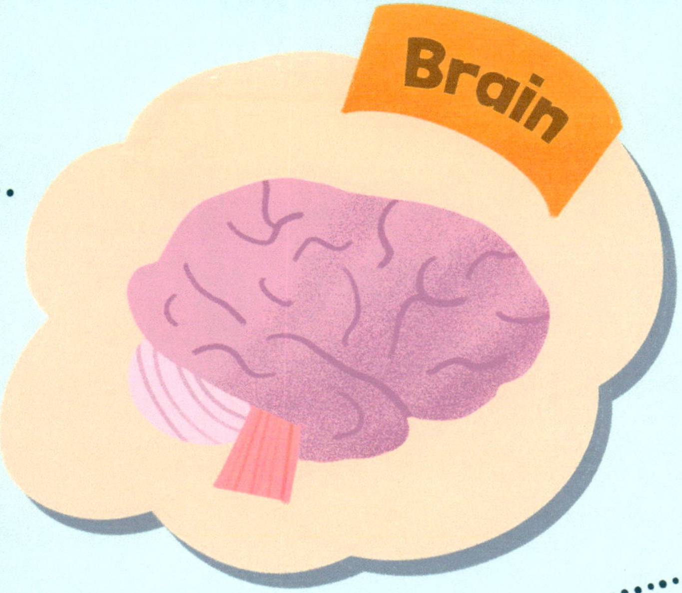

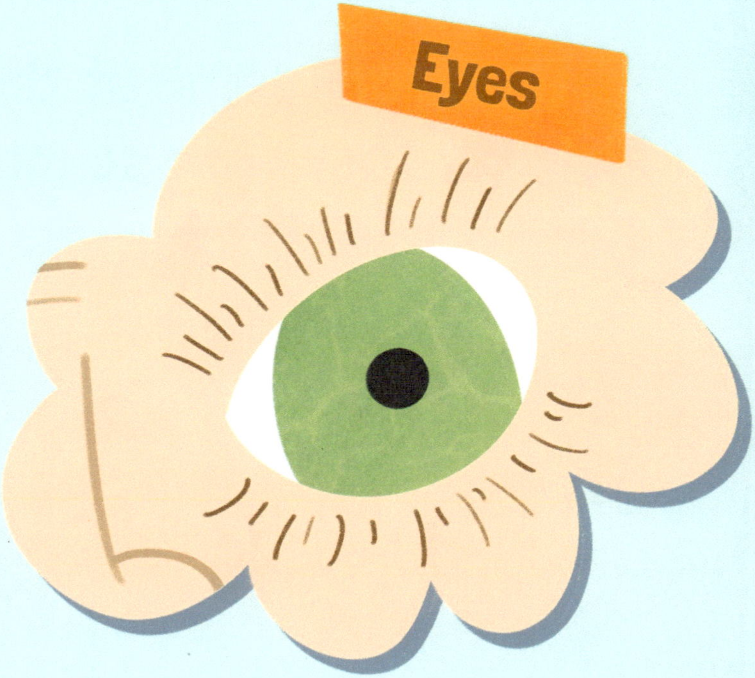

Anatomy is the study of the structures and parts of living things.

These are some of the branches in Biology. There are so many more. Do you have a favorite one you want to study?
Cell Biology
ZOOLOGY
Botany
EVOLUTION
Genetics
ANATOMY

Oh, they all sound exciting. I want to learn all about them.

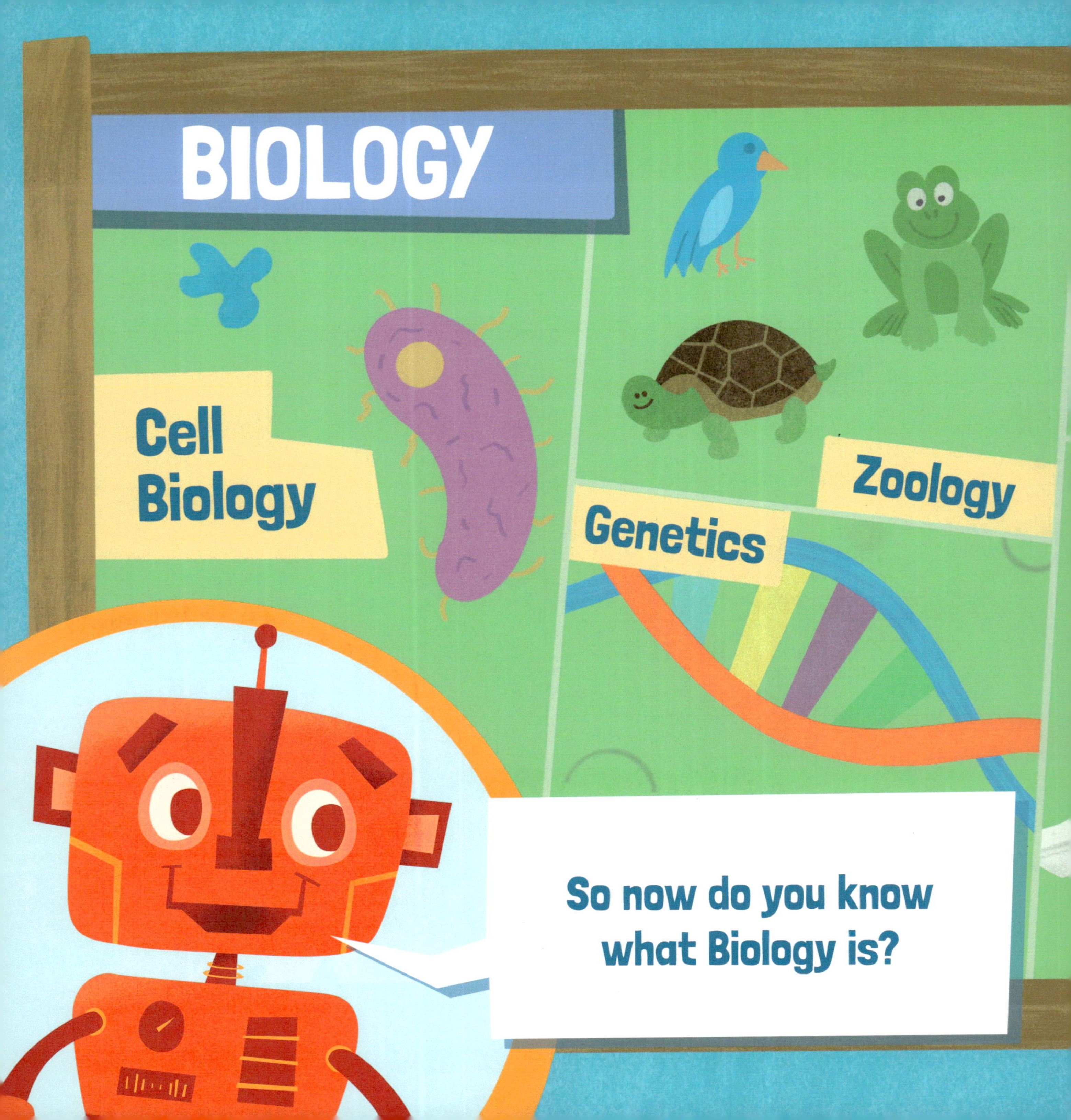
BIOLOGY
Cell Biology
Zoology
Genetics
So now do you know what Biology is?

Evolution
Botany
Anatomy
Yes. The study of living things! They include branches like Cell Biology, Zoology and Botany.

WHAT IS BIOLOGY?

For Curious Little Minds

Made in the USA
Las Vegas, NV
07 November 2026

57672509R00017